Multiple Choice Questions
in
Electrical Principles and Technology

Multiple Choice Questions

in

Electrical Principles and Technology

Advanced GNVQ

J. O. BIRD

NEWNES

Butterworth-Heinemann Ltd
Linacre House, Jordan Hill, Oxford OX2 8DP

A member of the Reed Elsevier plc group

OXFORD   LONDON   BOSTON
NEW DELHI   SINGAPORE   SYDNEY
TOKYO   TORONTO   WELLINGTON

First published 1996

© J. O. Bird 1996

British Library Cataloguing in Publication Data
A catalogue record for this book is available from the British Library

ISBN 0 7506 2667 4

Library of Congress Cataloguing in Publication Data
A catalogue record for this book is available from the Library of Congress

Printed in Great Britain by Martin's The Printers Ltd., Berwick upon Tweed

Contents

Preface

The two main aims of this resource book are, first, to provide a compact set of multiple choice items which lecturers might use to assess their students' progress, and secondly, to provide the student with practice in answering test papers similar to those presented by the Business and Technology Education Council and City and Guilds of London Institute for the GNVQ advanced mandatory units.

Each test paper comprises 20 questions and is designed to cover a wide range of topics. The tests are divided into three sections — Basic electrical engineering principles (which should be mainly revision), Electrical Principles and Electrical Technology (the latter two being optional units likely to be studied by most students on an electrical bias route).

A maximum of 1 hour is suggested for each test.

Each test paper, together with the answer grid (on page 45), is photocopiable.

With electrical principles and technology a large number of formulae are inherently involved; a list of formulae relevant to the two units covered is included as a helpful resource.

JOHN BIRD
Highbury College
Portsmouth

How to do multiple choice questions

In the multiple choice question papers for BTEC and C & G GNVQ modules, all of the questions are compulsory, so do not waste time initially reading through the paper. Begin at Question 1 and work steadily through the paper marking the answer grid with a tick in the box you think is the correct answer.

All of the questions in this book have only one correct answer, i.e. A, B, C or D.

If you come to a question you cannot answer make a sensible guess, mark the answer and come back to this question if you have time. You will not have marks deducted for wrong answers so you have nothing to lose by making an intelligent guess.

Test paper 1   Basic electrical engineering principles (1)

1 A resistance of 50 kΩ has a conductance of

   A 20 S
   B 0.02 S
   C 0.02 mS
   D 20 kS

2 Voltage drop is the

   A maximum potential
   B difference in potential between two points
   C voltage produced by a source
   D voltage at the end of a circuit

3 If two 4 Ω resistors are connected in parallel the effective resistance of the circuit is

   A 8 Ω
   B 4 Ω
   C 2 Ω
   D 1 Ω

4 Which of the following would apply to a moving-coil instrument?

   A An uneven scale, measuring d.c.
   B An even scale, measuring a.c.
   C An uneven scale, measuring a.c.
   D An even scale, measuring d.c.

5 In Question 4, which would refer to a moving-coil rectifier instrument?

6 Which of the following devices does not make use of the magnetic effect of an electric current?

   A an m.c. ammeter
   B a filament lamp
   C a relay
   D a transformer

7 For the current carrying conductor lying in the magnetic field shown, the direction of the force on the conductor is

   A to the left
   B upwards
   C to the right
   D downwards

(for Test 1, Question 7)

8 Which of the following statements is false?
The inductance of an inductor increases

   A with a short, thick coil
   B as the cross-sectional area of the coil decreases
   C as the number of turns increases
   D when wound on an iron core

9 The symbol for the unit of temperature coefficient of resistance is

   A $\Omega/°C$
   B Ω
   C $°C$
   D $\Omega/\Omega°C$

10 A current of 3 A flows for 5 h through a 100 Ω resistance.
The energy consumed by the resistance is

   A 4.5 kWh
   B 0.15 kWh
   C 1.5 kWh
   D 0.60 kWh

11 The effect of an air gap in a magnetic circuit is to

   A increase the reluctance
   B reduce the flux density

C   divide the flux
D   reduce the magnetomotive force

12 A voltmeter has an f.s.d. of 100 V, a sensitivity of 1 kΩ/V and an accuracy of ±2% of f.s.d. When the voltmeter is connected into a circuit it indicates 50 V. Which of the following statements is false?
A   Voltage reading is 50 ± 2 V.
B   Voltmeter resistance is 100 kΩ.
C   Voltage reading is 50 V $\pm 2\%$.
D   Voltage reading is 50 V $\pm 4\%$.

13 The unit of magnetic flux density is the
A   weber
B   weber per metre
C   ampere per metre
D   tesla

14 A current changing at a rate of 5 A/s in a coil of inductance 5 H induces an e.m.f. of
A   25 V in the opposite direction to the applied voltage
B   1 V in the same direction as the applied voltage
C   25 V in the same direction as the applied voltage
D   1 V in the opposite direction to the applied voltage

15 A 240 V, 60 W lamp has a working resistance of
A   1400 ohm
B   60 ohm
C   960 ohm
D   325 ohm

16 The capacitance of a capacitor is the ratio
A   charge to p.d. between plates
B   p.d. between plates to plate spacing

C   p.d. between plates to thickness of dielectric
D   p.d. between plates to charge

17 Two bar magnets are placed parallel to each other and about 2 cm apart, such that the south pole of one magnet is adjacent to the north pole of the other. With this arrangement, the magnets will
A   attract each other
B   have no effect on each other
C   repel each other
D   lose their magnetism

18 Five 2 V cells, each having an internal resistance of 0.2 Ω are connected in series to a load of resistance 14 Ω. The current flowing in the circuit is
A   10 A
B   1.4 A
C   1.5 A
D $\dfrac{2}{3}$ A

19 For the circuit of Question 18, the p.d. at the battery terminals is
A   10 V
B $9\dfrac{1}{3}$ V
C   0 V
D $10\dfrac{2}{3}$ V

20 What must be known in order to calculate the energy used by an electrical appliance?
A   voltage and current
B   current and time of operation
C   power and time of operation
D   current and resistance

Test paper 2 Basic electrical engineering principles (2)

1 The unit of quantity of electricity is the

  A volt
  B coulomb
  C ohm
  D joule

2 The equivalent resistance when a resistor of $\frac{1}{3}$ Ω is connected in parallel with a $\frac{1}{4}$ Ω resistor is

  A $\frac{1}{7}$ Ω

  B 7 Ω

  C $\frac{7}{12}$ Ω

  D $1\frac{5}{7}$ Ω

3 The charge on a 100 pF capacitor when the voltage applied to it is 2 kV is

  A 2 mC
  B 0.2 μC
  C 0.05 pC
  D 0.2 C

4 The p.d. applied to a 1 kΩ resistance in order that a current of 100 μA may flow is

  A 1 V
  B 100 V
  C 0.1 V
  D 10 V

5 The diagram shows a scale of a multi-range ammeter. What is the current indicated when switched to a 50 A scale?

  A 28 A
  B 5.6 A

  C 84 A
  D 8.4 A

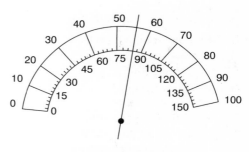

(for Test 2, Question 5)

6 Which of the following statements is false?

  A For non-magnetic materials reluctance is high.
  B Energy loss due to hysteresis is greater for harder magnetic materials than for softer magnetic materials.
  C The remanence of a ferrous material is measured in ampere/metre.
  D Absolute permeability is measured in henrys per metre.

7 Electromotive force is provided by

  A resistances
  B a conducting path
  C an electric current
  D an electrical supply source

8 The input and output currents of a system are 3 mA and 18 mA, respectively. The decibel current ratio of output to input current (assuming the input and load resistances are equal) is

  A 15.56
  B 6

C   1.6
D   7.78

9   The power dissipated by a resistor of 2 Ω when a current of 4 A passes through it is

A   1 W
B   8 W
C   32 W
D   16 W

10 An electric bell depends for its action on

A   a permanent magnet
B   reversal of current
C   a hammer and a gong
D   an electromagnet

11 The total flux in the core of an electrical machine is 20 mWb and its flux density is 1 T. The cross-sectional area of the core is

A   0.05 m^2
B   0.02 m^2
C   20 m^2
D   50 m^2

12 Capacitance is the ratio of

A   potential difference between plates to thickness of dielectric
B   number of lines of flux to area of field
C   charge to potential difference between plates
D   potential difference between plates to spacing

13 A conductor carries a current of 10 A at right angles to a magnetic field having a flux density of 500 mT. If the length of the conductor in the field is 20 cm, the force on the conductor is

A   100 kN
B   1 kN
C   100 N
D   1 N

14 Four 2 μF capacitors are connected in series. The equivalent capacitance is

A   8 μF
B   0.5 μF
C   2 μF
D   0.125 μF

15 A battery consists of

A   a cell
B   a circuit
C   a generator
D   a number of cells

16 A nickel coil has a resistance of 13 Ω at 50°C. If the temperature coefficient of resistance at 0°C is 0.006/°C, the resistance at 0°C is

A   16.9 Ω
B   10 Ω
C   43.3 Ω
D   0.1 Ω

17 The largest number of 100 W electric light bulbs which can be operated from a 240 V supply fitted with a 13 A fuse is

A   2
B   7
C   31
D   18

18 A strong permanent magnet is plunged into a coil and left in the coil. What is the effect produced on the coil after a short time?

A   There is no effect.
B   The insulation of the coil burns out.
C   A high voltage is induced.
D   The coil winding becomes hot.

19 Which of the following statements is true?

A   The capacity of a cell is measured in volts.

B   A primary cell converts electrical energy into chemical energy.

C   A positive electrode is termed the cathode.

D   Galvanizing iron helps to prevent corrosion.

20   The mutual inductance between two coils, when a current changing at 20 A/s in one coil induces an e.m.f. of 10 mV in the other, is

A   0.5 mH

B   200 mH

C   0.5 H

D   2 H

Test paper 3   Basic electrical engineering principles (3)

1   When a magnetic flux of 10 Wb links with a circuit of 20 turns in 2 s, the induced e.m.f. is

A   1 V
B   4 V
C   100 V
D   400 V

2   The coulomb is a unit of

A   power
B   voltage
C   energy
D   quantity of electricity

3   The power dissipated by a resistor of 10 Ω when a current of 2 A passes through it is

A   0.4 W
B   20 W
C   40 W
D   200 W

4   The diagram shows the poles X and Y of two bar magnets. The polarities of X and Y are

	X	Y
A	south	south
B	south	north
C	north	south
D	north	north

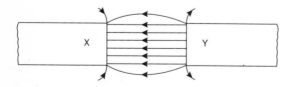

(for Test 3, Question 4)

5   The input and output powers of a system are 2 mW and 18 mW, respectively. The decibel power ratio of output power to input power is

A   9.54
B   9
C   1.9
D   19.08

6   A relay can be used to

A   decrease the current in a circuit
B   control a circuit more readily
C   increase the current in a circuit
D   control a circuit from a distance

7   Four 2 μF capacitors are connected in parallel.
The equivalent capacitance is

A   8 μF
B   0.5 μF
C   2 μF
D   0.125 μF

8   An e.m.f. of 1 V is induced in a 50 cm length of a conductor in a magnetic field of 0.4 T. The conductor is moving at a rate of

A   0.05 m/s
B   5 m/s
C   20 m/s
D   50 m/s

9   A charge of 240 C is transferred in 2 minutes. The current flowing is

A   120 A
B   480 A
C   2 A
D   8 A

10   A comparison can be made between mag-
     netic and electrical quantities. The equiv-
     alent electrical quantity to magnetic flux
     is

     A   resistivity
     B   current
     C   resistance
     D   e.m.f.

11   If two 4 Ω resistors are connected in
     series the effective resistance of the
     circuit is

     A   8 Ω
     B   4 Ω
     C   2 Ω
     D   1 Ω

12   There is a force of attraction between two
     current-carrying conductors when the cur-
     rent in them is

     A   of the same magnitude
     B   in the same direction
     C   of different magnitude
     D   in opposite directions

13   Which of the following statements is
     false?

     A   A Lechlanché cell is suitable for use
         in torches.
     B   A nickel-cadmium cell is an example
         of a primary cell.
     C   When a cell is being charged its ter-
         minal p.d. exceeds the cell e.m.f.
     D   A secondary cell may be recharged
         after use.

14   The energy stored in a 10 μF capacitor
     when charged to 500 V is

     A   1.25 J
     B   0.025 μJ
     C   1.25 mJ
     D   1.25 C

15   The terminal p.d. of a cell of e.m.f. 2 V
     and internal resistance of 0.2 Ω when sup-
     plying a current of 4 A will be

     A   2.8 V

     B   2 V
     C   1.2 V
     D   1.8 V

16   Which of the following is needed to
     extend the range of a milliammeter to
     read voltages of the order of 100 V?

     A   a parallel high-value resistance
     B   a series low-value resistance
     C   a parallel low-value resistance
     D   a series high-value resistance

17   Electrostatics is a branch of electricity
     concerned with

     A   energy flowing across a gap between
         conductors
     B   charges at rest
     C   charges in motion
     D   energy in the form of charges

18   The greater the internal resistance of a
     cell

     A   the greater the terminal p.d.
     B   the less the e.m.f.
     C   the greater the e.m.f.
     D   the less the terminal p.d.

19   The unit of resistivity is

     A   ohms
     B   ohm millimetre
     C   ohm metre
     D   ohm/metre

20   Which of the following statements is
     false?

     A   Fleming's left-hand rule or Lenz's
         law may be used to determine the
         direction of an induced e.m.f.
     B   An induced e.m.f. is set up whenever
         the magnetic field linking that circuit
         changes.
     C   The direction of an induced e.m.f. is
         always such as to oppose the effect
         producing it.
     D   The induced e.m.f. in any circuit is
         proportional to the rate of change of
         the magnetic flux linking the circuit.

Test paper 4   Basic electrical engineering principles (4)

1   A 10 Ω resistor is connected in parallel with a 15 Ω resistor and the combination is connected in series with a 12 Ω resistor.
The equivalent resistance of the circuit is
   A   18 Ω
   B   37 Ω
   C   27 Ω
   D   4 Ω

2   State which of the following is incorrect
   A $1 \text{ N} = 1 \text{ kg m/s}^2$
   B $1 \text{ V} = 1 \text{ J/C}$
   C $30 \text{ mA} = 0.03 \text{ A}$
   D $1 \text{ J} = 1 \text{ N/m}$

3   A bar magnet is moved at a steady speed of 1.0 m/s towards a coil of wire which is connected to a centre-zero galvanometer. The magnet is now withdrawn along the same path at 0.5 m/s. The deflection of the galvanometer is in the
   A   same direction as previously, with the magnitude of the deflection doubled
   B   opposite direction as previously, with the magnitude of the deflection halved
   C   same direction as previously, with the magnitude of the deflection halved
   D   opposite direction as previously, with the magnitude of the deflection doubled

4   In order that work may be done
   A   a supply of energy is required
   B   the circuit must have a switch
   C   coal must be burnt
   D   two wires are necessary

5   The diagram shows a rectangular coil of wire placed in a magnetic field and free to rotate about axis AB. If current flows into the coil at C, the coil will
   A   commence to rotate anticlockwise
   B   remain in the vertical position
   C   commence to rotate clockwise
   D   experience a force towards the north pole

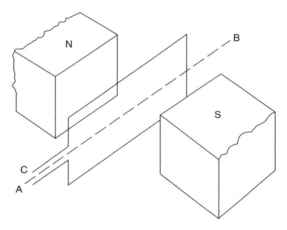

(for Test 4, Question 5)

6   Which of the following statements is true?
   A   Electric current is measured in volts.
   B   20 kΩ resistance is equivalent to 0.2 MΩ.
   C   An ammeter has a low resistance and must be connected in parallel with a circuit.
   D   An electrical insulator has a high resistance.

7   The magnetic field due to a current-carrying conductor takes the form of
   A   rectangles
   B   concentric circles

C   wavy lines
D   straight lines radiating outwards

8   The ohm is the unit of

A   charge
B   resistance
C   power
D   current

Questions 9 to 11 refer to the following data. A sinusoidal waveform is displayed on a c.r.o. screen. The peak-to-peak distance is 5 cm and the distance between cycles is 4 cm. The 'variable' switch is on 100 μs/cm and the 'volts/cm' switch is on 10 V/cm.

9   The maximum value of the waveform is

A   50 V
B   35.35 V
C   25 V
D   17.68 V

10   The frequency of the waveform is

A   2.5 Hz
B   25 Hz
C   250 Hz
D   2.5 kHz

11   The r.m.s. value of the waveform is

A   50 V
B   35.35 V
C   17.68 V
D   25 V

12   The energy used by a 1.5 kW heater in 5 minutes is

A   5 J
B   450 J
C   7500 J
D   450 000 J

13   The negative pole of a dry cell is made of

A   carbon
B   copper
C   zinc
D   mercury

14   A piece of graphite has a cross-sectional area of 10 mm^2. If its resistance is 0.1 Ω and its resistivity 10×10^{-8} Ωm, its length is

A   10 m
B   10 cm
C   10 mm
D   10 km

15   The p.d. across a 10 μF capacitor to charge it with 10 mC is

A   100 V
B   1 kV
C   1 V
D   10 V

16   A current of 10 A in a coil of 1000 turns produces a flux of 10 mWb linking with the coil. The coil inductance is

A   1 H
B 10^6 H
C   100 H
D   1 mH

17   Which of the following statements is false?
When two metal electrodes are used in a simple cell, the one that is higher in the electrochemical series

A   tends to dissolve in the electrolyte
B   is always the negative electrode
C   reacts most readily with oxygen
D   acts as the anode

18   The effect of connecting an additional parallel load to an electrical supply source is to increase the

A   resistance of the load
B   voltage of the source
C   current taken from the source
D   p.d. across the load

19   The force on an electron travelling at 10^7 m/s in a magnetic field of density 10 μT is 1.6×10^{-17} N. The electron has a charge of

A 1.6×10^{-28} C
B 1.6×10^{-15} C

C 1.6×10^{-19} C

D 1.6×10^{-25} C

20 Which of the following statements is false?

A An electrolytic capacitor must be used only on a.c. supplies.

B A paper capacitor generally has a shorter service life than most other types of capacitor.

C An air capacitor is normally a variable type.

D Plastic capacitors generally operate satisfactorily under conditions of high temperature.

Test paper 5 Basic electrical engineering principles (5)

1 When three 3 Ω resistors are connected in parallel, the total resistance is

   A 0.333 Ω
   B 9 Ω
   C 1 Ω
   D 3 Ω

2 Which of the following formulae for electrical power is incorrect?

   A VI

   B $\dfrac{V}{I}$

   C I^2R

   D $\dfrac{V^2}{R}$

3 A coil of wire has a resistance of 10 Ω at 0°C. If the temperature coefficient of resistance for the wire is 0.004/°C, its resistance at 100°C is

   A 14 Ω
   B 1.4 Ω
   C 0.4 Ω
   D 10 Ω

4 With the switch in the circuit shown closed, the ammeter reading will indicate

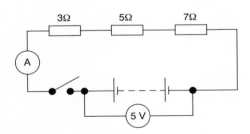

(for Test 5, Question 4)

   A $1\dfrac{2}{3}$ A

   B 75 A

   C $\dfrac{1}{3}$ A

   D 3 A

5 The capacitance of a variable air capacitor is at maximum when

   A the movable plates half overlap the fixed plates
   B the movable plates are most widely separated from the fixed plates
   C the movable plates are closer to one side of the fixed plate than to the other
   D both sets of plates are exactly meshed

6 When an atom loses an electron, the atom

   A becomes positively charged
   B disintegrates
   C experiences no effect at all
   D becomes negatively charged

7 A 230 V, 100 W lamp has a working resistance of

   A 2.3 Ω
   B 100 Ω
   C 529 Ω
   D 23 kΩ

Questions 8 to 10 refer to the diagram of double beam c.r.o. waveform traces shown.

8 The amplitude of waveform P is

   A 75 V
   B 50 V
   C 30 V
   D 15 V

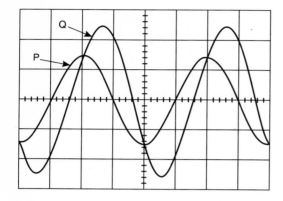

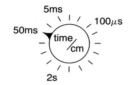

(for Test 5, Questions 8–10)

9   The peak-to-peak value of waveform Q is
    A   75 V
    B   50 V
    C   30 V
    D   15 V

10 The periodic time of waveform P is
    A   100 μs
    B   40 μs
    C   0.1 s
    D   0.2 s

A comparison can be made between magnetic and electrical quantities.
Questions 11 and 12 refer to this comparison.

11 The equivalent magnetic quantity to electrical resistance is
    A   flux
    B   m.m.f.
    C   reluctance
    D   relative permeability

12 The equivalent electrical quantity to magnetomotive force is
    A   resistivity

B   e.m.f.
C   resistance
D   current

13 The total resistance of two resistors R_1 and R_2 when connected in parallel is given by

    A $R_1 + R_2$

    B $\dfrac{1}{R_1} + \dfrac{1}{R_2}$

    C $\dfrac{R_1 + R_2}{R_1 R_2}$

    D $\dfrac{R_1 R_2}{R_1 + R_2}$

14 When a current-carrying conductor is placed in and at right angles to a magnetic field
    A   nothing happens to the field or the conductor
    B   the conductor current ceases to flow
    C   the current-carrying conductor experiences a force
    D   the magnetic field collapses

15 The terminal p.d. of a cell of e.m.f. 2 V and internal resistance 0.1 Ω when supplying a current of 5 A will be
    A   1.5 V
    B   2 V
    C   1.9 V
    D   2.5 V

16 A current of 3 A flows for 50 h through a 6 Ω resistor. The energy consumed by the resistor is
    A   0.9 kWh
    B   2.7 kWh
    C   9 kWh
    D   27 kWh

17 Self-inductance occurs when
    A   the current is changing
    B   the circuit is changing
    C   the flux is changing
    D   the resistance is changing

18   If a conductor is horizontal, the current flowing from left to right and the direction of the surrounding magnetic field is from above to below, the force exerted on the conductor is

   A   from left to right
   B   from below to above
   C   towards the viewer
   D   away from the viewer

19   The resistance of a 2 km length of cable of cross-sectional area 2 mm^2 and resistivity of 2×10^{-8} Ωm is

   A   0.02 Ω
   B   20 Ω
   C   0.02 mΩ
   D   200 Ω

20   The current flowing in a 500 turn coil wound on an iron ring is 4 A. The reluctance of the circuit is 2×10^6/H. The flux produced is

   A   1 mWb
   B   1000 Wb
   C   1 Wb
   D   62.5 μWb

Test paper 6 Basic electrical engineering principles (6)

1 The unit of current is

   A   the volt
   B   the coulomb
   C   the joule
   D   the ampere

2 With the switch in the circuit shown closed, the ammeter reading will indicate

   A   108 A

   B $\dfrac{1}{3}$ A

   C   3 A

   D $4\dfrac{3}{5}$ A

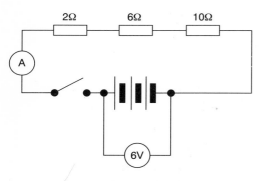

(for Test 6, Question 2)

3 A 6 Ω resistor is connected in parallel with the three resistors of the circuit shown above. With the switch closed the ammeter reading will indicate

   A $\dfrac{3}{4}$ A

   B   4 A

   C $\dfrac{1}{4}$ A

   D $1\dfrac{1}{3}$ A

4 The charge on a 10 pF capacitor when the voltage applied to it is 10 kV is

   A   100 μC
   B   0.1 C
   C   0.1 μC
   D   0.01 μC

5 State which of the following is false.
The capacitance of a capacitor

   A   is proportional to the distance between the plates
   B   is proportional to the cross-sectional area of the plates
   C   depends on the number of plates
   D   is proportional to the relative permittivity of the dielectric

6 The current which flows when 0.1 coulomb is transferred in 10 ms is

   A   1 A
   B   10 A
   C   10 mA
   D   100 mA

7 The input and output voltages of a system are 500 μV and 500 mV, respectively. The decibel voltage ratio of output to input voltage (assuming input resistance equals load resistance) is

   A   1000
   B   30
   C   0
   D   60

8   Faraday's laws of electromagnetic induction are related to

    A   the e.m.f. of a generator
    B   the e.m.f. of a chemical cell
    C   the current flowing in a conductor
    D   the strength of a magnetic field

9   A current of 2 A flows for 10 h through a 100 Ω resistor. The energy consumed by the resistor is

    A   0.5 kWh
    B   4 kWh
    C   2 kWh
    D   0.02 kWh

10   Which of the following statements is false?

    A   The Schering bridge is normally used for measuring unknown capacitances.
    B   A.c. electronic measuring instruments can handle a much wider range of frequency than the moving-coil instrument.
    C   A complex waveform is one which is non-sinusoidal.
    D   A square wave normally contains the fundamental and even harmonics.

11   The length of a certain conductor of resistance 100 Ω is doubled and its cross-sectional area is halved. Its new resistance is

    A   400 Ω
    B   200 Ω
    C   50 Ω
    D   100 Ω

Questions 12 to 16 refer to the following data. A coil of 100 turns is wound uniformly on a wooden ring. The ring has a mean circumference of 1 m and a uniform cross-sectional area of 10 cm^2. The current in the coil is 1 A.

12   The magnetomotive force is

    A   1 A
    B   10 A
    C   100 A
    D   1000 A

13   The magnetic field strength is

    A   1 A/m
    B   10 A/m
    C   100 A/m
    D   1000 A/m

14   The magnetic flux density is

    A   100 T
    B 8.85×10^{-10} T
    C $4\pi \times 10^{-7}$ T
    D 40π μT

15   The magnetic flux is

    A 0.04π μWb
    B   0.01 Wb
    C   8.85 μWb
    D 4π μWb

16   The reluctance is

    A $\dfrac{10^8}{4\pi}$ H^{-1}

    B   1000 H^{-1}

    C $\dfrac{2.5}{\pi} \times 10^9$ H^{-1}

    D $\dfrac{10^8}{8.85}$ H^{-1}

17   A potentiometer is used to

    A   compare voltages
    B   compare currents
    C   measure power factor
    D   measure phase sequence

18   An e.m.f. of 1 V is induced in a conductor moving at 10 cm/s in a magnetic field of 0.5 T. The effective length of the conductor in the magnetic field is

    A   20 cm
    B   5 m
    C   20 m
    D   50 m

19 Five cells, each with an e.m.f. of 2 V and internal resistance 0.5 Ω, are connected in series. The resulting battery will have

A   an e.m.f. of 2 V and an internal resistance of 0.5 Ω

B   an e.m.f. of 10 V and an internal resistance of 2.5 Ω

C   an e.m.f. of 2 V and an internal resistance of 0.1 Ω

D   an e.m.f. of 10 V and an internal resistance of 0.1 Ω

20 If the 5 cells of Question 19 are connected in parallel the resulting battery will have

A   an e.m.f. of 2 V and an internal resistance of 0.1 Ω

B   an e.m.f. of 10 V and an internal resistance of 2.5 Ω

C   an e.m.f. of 2 V and an internal resistance of 0.5 Ω

D   an e.m.f. of 10 V and an internal resistance of 0.1 Ω

Test paper 7 Basic electrical engineering principles (7)

1 The unit of flux density is the

A henry
B weber
C tesla
D farad

2 If in the circuit shown, the reading on the voltmeter is 5 V and the reading on the ammeter is 25 mA, the resistance of resistor R is

A 0.005 Ω
B 5 Ω
C 125 Ω
D 200 Ω

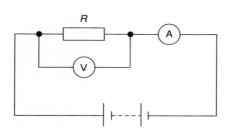

(for Test 7, Question 2)

3 The force F newtons on a conductor of length l metres carrying a current I amperes perpendicular to a magnetic field of flux density B teslas is given by

A $\dfrac{BI}{l}$

B $\dfrac{B}{Il}$

C BIl

D $\dfrac{Bl}{I}$

4 The terminal p.d. of a 2 V cell is 1.4 V when supplying a current of 6 A. The internal resistance of the cell is

A 0.1 Ω
B 0.57 Ω
C 3.6 Ω
D 20.4 Ω

5 When the Wheatstone bridge shown is balanced

A $R_1R_2 = R_3R_4$
B $R_1 + R_2 = R_3 + R_4$
C $R_1 + R_3 = R_2 + R_4$
D $R_1R_4 = R_2R_3$

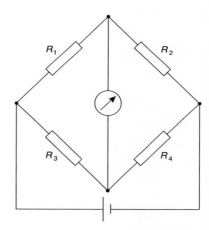

(for Test 7, Question 5)

6 When a voltage of 1 kV is applied to a capacitor, the charge on the capacitor is 500 nF. The capacitance of the capacitor is

A 2×10^9 F
B 0.5 nF
C 0.5 mF
D 0.5 pF

7   For the hysteresis loop shown, the length
    Ox represents

    A   remanence
    B   reluctance
    C   coercive force
    D   reactance

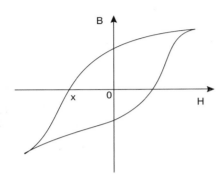

(for Test 7, Question 7)

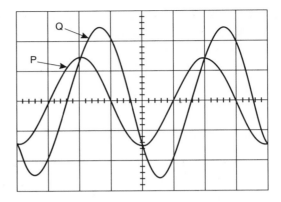

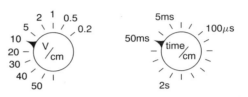

(for Test 7, Questions 10–12)

8   A conductance of 20 mS is equivalent to
    a resistance of

    A   50 kΩ
    B   0.05 Ω
    C   50 Ω
    D   0.05 mΩ

9   The unit of electric flux density is the

    A   weber
    B   weber per square metre
    C   coulombs per ampere
    D   coulombs per square metre

Double-beam c.r.o. waveform traces are shown
in the diagram.
Questions 10 to 12 refer to these waveforms.

10 The frequency of waveform Q is

    A   5 Hz
    B   25 Hz
    C   10 kHz
    D   25 kHz

11 The r.m.s. value of waveform P is

    A $\dfrac{15}{\sqrt{2}}$ V

    B $\dfrac{25}{\sqrt{2}}$ V

    C   15 V

    D $\dfrac{50}{\sqrt{2}}$ V

12 The phase displacement of waveform Q
    relative to waveform P is

    A   54° leading

    B $\dfrac{3\pi}{10}$ rad lagging

    C   108° lagging

    D $\dfrac{3\pi}{10}$ rad leading

13 The current flowing through a wire-
    wound resistor can be decreased by

    A   using a wire of lower resistivity
    B   increasing the cross-sectional area of
        the wire
    C   increasing the length of the wire used
        to wind the resistor
    D   cooling the wire to a lower tempera-
        ture

14 A current flows for 4 hours through a 200 Ω resistance. If the e.m.f. across the resistance is 400 V, the energy consumed by the resistance is

A   3.2 kWh
B   20 kWh
C   320 kWh
D   0.40 kWh

15 When a current I amperes flows through a resistance R ohms, the power developed, in watts, is

A IR^2

B I^2R

C $\dfrac{I^2}{R}$

D $\dfrac{I}{R^2}$

16 When two resistors of $\dfrac{1}{4}\ \Omega$ and $\dfrac{1}{2}\ \Omega$ are connected in parallel their effective resistance is

A   6 Ω

B   2 Ω

C $\dfrac{3}{4}\ \Omega$

D $\dfrac{1}{6}\ \Omega$

17 Which of the following units for the quantities listed below is incorrect?

A   dielectric strength V/m
B   resistivity Ω

C   reluctance H^{-1}
D   temperature coefficient of resistance $\Omega/\Omega°C$

18 The supply e.m.f. E in the circuit shown is

A   26 V
B   18 V
C   14 V
D   8 V

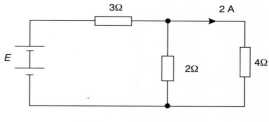

(for Test 7, Question 18)

19 The energy of a secondary cell is usually renewed

A   by heating it
B   it cannot be renewed at all
C   by renewing its chemical
D   by passing a current through it

20 If the total flux in a magnetic circuit is 2 mWb and the cross-sectional area of the circuit is 10 cm^2, the flux density is

A   0.2 T
B   2 T
C   20 T
D   20 mT

Test paper 8   Electrical principles (1)

1   Which of the following statements is true? For the junction in the network shown

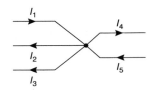

(for Test 8, Question 1)

A $I_1 - I_2 - I_3 - I_4 + I_5 = 0$
B $I_1 + I_2 + I_3 = I_4 + I_5$
C $I_2 + I_3 + I_5 = I_1 + I_4$
D $I_5 - I_4 = I_3 - I_2 + I_1$

2   An inductance of 10 mH connected across a 100 V, 50 Hz supply has an inductive reactance of

A 10π Ω
B 1000π Ω
C π Ω
D π MΩ

3   In the diagram, at the instant shown, the generated e.m.f. will be

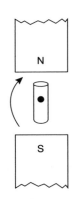

(for Test 8, Question 3)

A   a maximum value
B   an r.m.s. value
C   an average value
D   zero

4   A rectifier conducts

A   direct currents in one direction
B   alternating currents in both directions
C   direct currents in both directions
D   alternating currents in one direction

5   The time constant for a circuit containing a capacitance of 100 nF in series with a 5 Ω resistance is

A   0.5 μs
B   20 ns
C   5 μs
D   50 μs

6   For the circuit shown, voltage V is

A   0 V
B   20 V
C   4 V
D   16 V

7   For the circuit shown below, current I_1 is

A   25 A

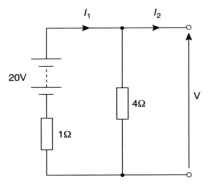

(for Test 8, Questions 6–8)

   B   4 A
   C   0 A
   D   20 A

8   For the circuit shown above, current I_2 is

   A   25 A
   B   4 A
   C   0 A
   D   20 A

9   The value of an alternating current at any given instant is

   A   a maximum value
   B   a peak value
   C   an instantaneous value
   D   an r.m.s. value

10 The impedance of a coil, which has a resistance of X ohms and an inductance of Y henrys, connected across a supply of frequency K Hz, is

   A $2\pi KY$
   B $X + Y$
   C $\sqrt{(X^2 + Y^2)}$
   D $\sqrt{\{X^2 + (2\pi KY)^2\}}$

11 In Question 10, the phase angle between the current and the applied voltage is given by

   A $\arctan \dfrac{2\pi KY}{X}$

   B $\arctan \dfrac{Y}{X}$

   C $\arctan \dfrac{X}{2\pi KY}$

   D $\tan \dfrac{2\pi KY}{X}$

12 The amplitude of the current I flowing in the circuit shown is

   A   21 A
   B   16.8 A
   C   28 A
   D   12 A

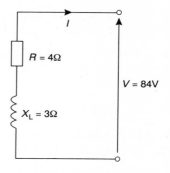

(for Test 8, Question 12)

Questions 13 to 20 refer to the following information.

A two-branch parallel circuit containing a 10 Ω resistance in one branch and a 100 μF capacitor in the other, has a 120 V, $\dfrac{2}{3}\pi$ kHz supply connected across it.

13 The current flowing in the resistance is

   A   83.3 mA
   B   1.2 A
   C   12 A
   D   1200 A

14 The capacitive reactance of the capacitor is

   A   7.5 kΩ
   B   7.5 Ω
   C   133 Ω
   D   0.133 Ω

15 The current flowing in the capacitor is

   A   902 A
   B   16 A
   C   16 mA
   D   0.902 A

16 The supply current is

   A   4 A
   B   10.58 A
   C   20 A
   D   28 A

17 The supply phase angle is

   A $\arctan \dfrac{4}{3}$ leading

B   arctan $\dfrac{4}{3}$ lagging

C   arctan $\dfrac{3}{4}$ leading

D   arctan $\dfrac{3}{4}$ lagging

18   The circuit impedance is

    A   2.5 Ω

    B   6 Ω

    C   6.61 Ω

    D   17.5 Ω

19   The power consumed by the circuit is

    A   360 W

    B   720 W

    C   1.92 kW

    D   1.44 kW

20   The power factor of the circuit is

    A   0.6 lagging

    B   0.8 leading

    C   0.8 lagging

    D   0.6 leading

Test paper 9 Electrical principles (2)

1 An alternating current completes 100 cycles in 0.1 s. Its frequency is

  A 20 Hz
  B 100 Hz
  C 0.002 Hz
  D 1 kHz

2 For the circuit shown, the internal resistance, r, is given by

  A $\dfrac{I}{V-E}$

  B $\dfrac{V-E}{I}$

  C $\dfrac{I}{E-V}$

  D $\dfrac{E-V}{I}$

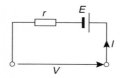

(for Test 9, Question 2)

3 The time constant for a circuit containing an inductance of 100 mH in series with a resistance of 4 Ω is

  A 25 ms
  B 400 s
  C 0.4 s
  D 40 s

4 Which of the following statements is false?

  A It is cheaper to use a.c. than d.c.

  B Distribution of a.c. is more convenient than with d.c. since voltages may be readily altered using transformers.

  C An alternator is an a.c. generator.

  D A rectifier changes d.c. into a.c.

5 The maximum power transfer from the source to the load R_L in the circuit shown is

  A 96 W
  B 72 W
  C 192 W
  D 48 W

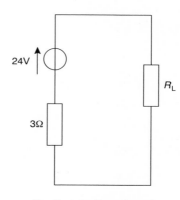

(for Test 9, Question 5)

6 The current flowing in the branches of a d.c. circuit may be determined using

  A Faraday's laws
  B Lenz's laws
  C Kirchhoff's laws
  D Fleming's left-hand rule

7 The graph shown represents the growth of current in an $L-R$ series circuit connected

to a d.c. voltage V volts. The equation for the graph is

A $i = I(1 - e^{-Rt/L})$
B $i = Ie^{-Li/t}$
C $i = Ie^{-Rt/L}$
D $i = I(1 - e^{RL/t})$

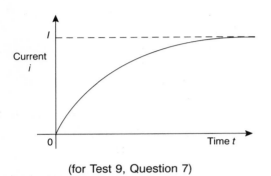

(for Test 9, Question 7)

8   The equivalent resistance across terminals AB of the circuit shown is

A   9.31 Ω
B   7.24 Ω
C   10.0 Ω
D   6.75 Ω

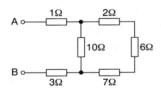

(for Test 9, Question 8)

9   Which of the following statements is false?

A   The Q-factor at resonance in a parallel circuit is the voltage magnification.
B   Improving power factor reduces the current flowing through a system.
C   The supply current is a minimum at resonance in a parallel circuit.
D   The circuit impedance is a maximum at resonance in a parallel circuit.

In Questions 10 and 11, a series connected C-R circuit is suddenly connected to a d.c. source of V volts. Which of the statements is false?

10 A   The initial current flowing is given by $\dfrac{V}{R}$.
B   The current grows exponentially.
C   The time constant of the circuit is given by CR.
D   The final value of the current is zero.

11 A   The capacitor voltage is equal to the voltage drop across the resistor.
B   The voltage drop across the resistor decays exponentially.
C   The initial capacitor voltage is zero.
D   The initial voltage drop across the resistor is IR, where I is the steady-state current.

12 The value of the voltage V in the circuit shown is

A $119 \underline{/22.62°}$ V
B $91 \underline{/45.24°}$ V
C $91 \underline{/0°}$ V
D $91 \underline{/22.62°}$ V

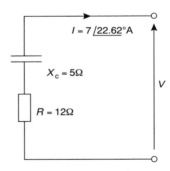

(for Test 9, Question 12)

13 The open-circuit voltage E across terminals XY of the circuit shown is

A   0 V
B   16 V
C   4 V
D   20 V

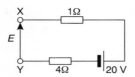

(for Test 9, Question 13)

14 The number of complete cycles of an alternating current occurring in one second is known as

A   the maximum value of the alternating current

B   the frequency of the alternating current

C   the peak value of the alternating current

D   the r.m.s. or effective value

15 A series a.c. circuit comprising a coil of inductance 100 mH and resistance 1 Ω and a 10 μF capacitor is connected across a 10 V supply. At resonance the p.d. across the capacitor is

A   10 kV

B   10 V

C   100 V

D   1 kV

Questions 16 to 18 refer to the circuit shown. Values are stated correct to 2 significant figures.

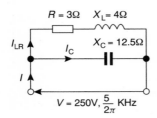

(for Test 9, Questions 16–18)

16 Which of the following statements is false?

A   The impedance of the R–L branch is 5 Ω.

B   The inductance of the R–L branch is 0.80 H.

C   The current I_{LR} is 50 A.

D   The current in the capacitive branch is 20 A.

17 Which of the following statements is false?

A   The capacitance of the capacitive branch is 16 μF.

B   The 'in-phase' component of the supply current is 30 A.

C   The 'quadrature' component of the supply current is 40 A.

D   The supply current I is 36 A.

18 Which of the following statements is false?

A   The circuit phase angle is 33.69° leading.

B   The circuit impedance is 6.9 Ω.

C   The circuit power factor is 0.83 lagging.

D   The reactive power is 5.0 kvar.

19 An a.c. supply is 70.7 V, 50 Hz. Which of the following statements is false?

A   The periodic time is 20 ms.

B   The peak value of the voltage is 70.7 V.

C   The r.m.s. value of the voltage is 70.7 V.

D   The peak value of the voltage is 100 V.

20 When a capacitor is connected to an a.c. supply the current

A   leads the voltage by 180°

B   is in phase with the voltage

C   leads the voltage by $\dfrac{\pi}{2}$ rad

D   lags the voltage by 90°

Test paper 10   Electrical principles (3)

1   For the circuit shown, voltage V is

   A   12 V
   B   2 V
   C   10 V
   D   0 V

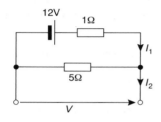

(for Test 10, Questions 1–3)

2   For the circuit shown above, current I_1 is

   A   2 A
   B   14.4 A
   C   0.5 A
   D   0 A

3   For the circuit shown above, current I_2 is

   A   2 A
   B   14.4 A
   C   0.5 A
   D   0 A

4   At the resonant frequency, which of the following statements is true for the parallel a.c. circuit shown in the following diagram?

   A I_C lags I_L by 90°.
   B I is in phase with V.
   C I_C is a minimum value.
   D I_L leads I by 90°.

5   If the supply frequency is increased in a series $R–L–C$ circuit at resonance and the

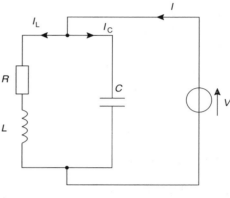

(for Test 10, Question 4)

values of L, C and R are constant, the circuit will become

   A   capacitive
   B   resistive
   C   inductive
   D   resonant

6   A capacitor which is charged to V volts is discharged through a resistor of R ohms. Which of the following statements is false?

   A   The initial current flowing is $\dfrac{V}{R}$ amperes.

   B   The voltage drop across the resistor is equal to the capacitor voltage.

   C   The time constant of the circuit is CR seconds.

   D   The current grows exponentially to a final value of $\dfrac{V}{R}$ amperes.

7   In a series a.c. circuit the voltage across a pure inductance is 12 V and the voltage across a pure resistance is 5 V. The supply voltage is

A   13 V
B   17 V
C   7 V
D   2.4 V

8   The value normally stated when referring to alternating currents and voltages is the

A   instantaneous value
B   r.m.s. value
C   average value
D   peak value

9   A two-branch parallel circuit consists of a 15 mH inductance in one branch and a 50 μF capacitor in the other across a 120 V, $\frac{1}{\pi}$ kHz supply.

The supply current is

A   8 A leading by $\frac{\pi}{2}$ rad
B   16 A lagging by 90°
C   8 A lagging by 90°
D   16 A leading by $\frac{\pi}{2}$ rad

10   The supply of electrical energy for a consumer is usually by a.c. because

A   it is most suitable for variable speed motors
B   transmission and distribution are more easily effected
C   the volt drop in cables is minimal
D   cable power losses are negligible

11   For the circuit shown, the value of Q-factor is

A   50
B   100

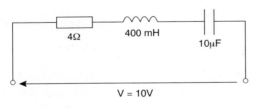

V = 10V

(for Test 10, Question 11)

C 5×10^{-4}
D   40

12   An alternating voltage is given by $v = 100 \sin(50\pi t - 0.30)$ V. Which of the following statements is true?

A   The r.m.s. voltage is 100 V.
B   The periodic time is 20 ms.
C   The frequency is 25 Hz.
D   The voltage is leading $v = 100 \sin 50\pi t$ by 0.30 radians.

13   An LR–C parallel circuit has the following component values: $R = 10$ Ω, $L = 10$ mH, $C = 10$ μF, $V = 100$ V. Which of the following statements is false?

A   The resonant frequency f_r is $\frac{1.5}{\pi}$ kHz.
B   The current at resonance is 1 A.
C   The dynamic resistance is 100 Ω.
D   The circuit Q-factor at resonance is 30.

14   Which of the following statements is false?

A   Impedance is at a minimum at resonance in an a.c. series circuit.
B   The product of r.m.s. current and voltage gives the apparent power in an a.c. circuit.
C   Current is at a maximum at resonance in an a.c. series circuit.
D $\frac{\text{Apparent power}}{\text{True power}}$ gives power factor.

15   For the circuit shown on the following page, maximum power transfer from the source is required.
For this to be so, which of the following statements is true?

A $R_2 = 10$ Ω
B $R_2 = 30$ Ω
C $R_2 = 7.5$ Ω
D $R_2 = 15$ Ω

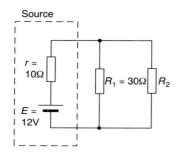

(for Test 10, Question 15)

Questions 16 to 20 refer to the following data. An inductor of inductance 0.1 H and negligible resistance is connected in series with a 50 Ω resistance to a 20 V d.c. supply.

16   The value of the time constant of the circuit is

A   500 s
B   2 ms
C   5 ms
D   5 s

17   The final value of current flowing in the circuit is

A   0
B   0.4 A

C   1 mA
D   2.5 A

18   The approximate value of the voltage across the resistor at a time equal to the time constant after being connected to the supply is

A   0
B   7.4 V
C   12.6 V
D   20 V

19   The initial value of the voltage across the inductor is

A   20 V
B   18.4 V
C   12.6 V
D   0

20   The final value of the steady-state voltage across the inductor is

A   20 V
B   18.4 V
C   12.6 V
D   0

Test paper 11   Electrical principles (4)

1 Which of the following statements is true?
   For the circuit shown below

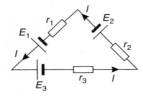

(for Test 11, Question 1)

   A $E_1 + E_2 + E_3 = Ir_1 + Ir_2 + Ir_3$
   B $E_2 + E_3 - E_1 - I(r_1 + r_2 + r_3) = 0$
   C $I(r_1 + r_2 + r_3) = E_1 - E_2 - E_3$
   D $E_2 + E_3 - E_1 = Ir_1 + Ir_2 + Ir_3$

2 An alternating current is given by
   $i = 70.71 \sin(100\pi t + 0.25)$ mA.
   Which of the following statements is true?

   A The r.m.s. current is 70.71 mA.
   B The current is lagging $70.71 \sin 100\pi t$ by 14.32°.
   C The periodic time is 100π ms.
   D The frequency is 50 Hz.

3 A capacitor of 1 μF is connected to a 50 Hz supply. The capacitive reactance is

   A 50 MΩ

   B $\dfrac{10}{\pi}$ kΩ

   C $\dfrac{\pi}{10^4}$ Ω

   D $\dfrac{10}{\pi}$ Ω

4 The maximum power transferred by the source in the circuit shown is
   A 5 W
   B 50 W
   C 40 W
   D 200 W

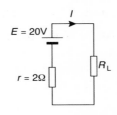

(for Test 11, Question 4)

5 Inductive reactance results in a current that

   A leads the voltage by 90°
   B is in phase with the voltage
   C leads the voltage by π rad

   D lags the voltage by $\dfrac{\pi}{2}$ rad

6 A series *RLC* circuit has a resistance of 8 Ω, an inductance of 100 mH and a capacitance of 5 μF. If the current flowing is 2 A, the impedance at resonance is

   A 160 Ω
   B 8 Ω
   C 8 mΩ
   D 16 Ω

7 With reference to the circuit shown below, which of the following statements is correct?

   A $V_{PQ} = 15$ V
   B $V_{PQ} = 2$ V

C   When a load is connected between P and Q, current would flow from Q to P.

D $V_{PQ} = 20$ V

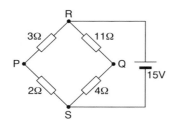

(for Test 11, Question 7 and 8)

8   In the above circuit, if the 15 V battery is replaced by a short-circuit, the equivalent resistance across terminal PQ is

A   4.13 Ω
B   4.20 Ω
C   20 Ω
D   4.29 Ω

9   State which of the following is false? For a sine wave

A   the peak factor is 1.414
B   the r.m.s. value is 0.707 × peak value
C   the average value is 0.637 × r.m.s. value
D   the form factor is 1.11

10   In an R–L–C series a.c. circuit a current of 5 A flows when the supply voltage is 100 V. The phase angle between current and voltage is 60° lagging. Which of the following statements is false?

A   The equivalent circuit reactance is 20 Ω.
B   The circuit is effectively inductive.
C   The apparent power is 500 VA.
D   The true power is 250 W.

11   The magnitude of the impedance of the circuit shown is

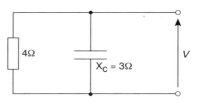

(for Test 11, Question 11)

A   7 Ω
B   5 Ω
C   2.4 Ω
D   1.71 Ω

12   An alternating voltage of maximum value 100 V is applied to a lamp. Which of the following direct voltages, if applied to the lamp, would cause the lamp to light with the same brilliance?

A   70.7 V
B   100 V
C   63.7 V
D   141.4 V

13   In the circuit shown, the magnitude of the supply current I is

A   17 A
B   7 A
C   15 A
D   23 A

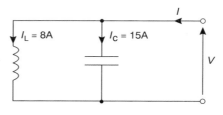

(for Test 11, Question 13)

Questions 14 to 20 refer to the following information.
An uncharged 2 μF capacitor is connected in series with a 5 MΩ resistor to a 100 V, constant voltage, d.c. supply.

14   The time constant of the circuit is

A   0.4 s
B   1 s
C   10 s
D   2.5 s

15   The final voltage across the capacitor is

A   0 V
B   10 V
C   50 V
D   100 V

16   The initial voltage across the resistor is

A   0 V
B   10 V
C   50 V
D   100 V

17   The final voltage across the resistor is

A   0 V
B   10 V

C   50 V
D   100 V

18   The initial voltage across the capacitor is

A   0 V
B   10 V
C   50 V
D   100 V

19   The initial current flowing in the circuit is

A   500 A
B   20 A
C   20 μA
D   0 A

20   The final current flowing in the circuit is

A   500 A
B   20 A
C   20 μA
D   0 A

Test paper 12 Electrical technology (1)

Questions 1 to 5 refer to the following information.

Three loads, each of 10 Ω resistance, are connected in star to a 415 V, three phase supply.

1 The line voltage is

    A   240 V
    B   293 V
    C   415 V
    D   719 V

2 The phase voltage is

    A   240 V
    B   293 V
    C   415 V
    D   719 V

3 The phase current is

    A   72 A
    B   42 A
    C   29 A
    D   24 A

4 The line current is

    A   72 A
    B   42 A
    C   29 A
    D   24 A

5 The total power dissipated in the load is

    A   51.67 kW
    B   29.83 kW
    C   17.22 kW
    D   9.94 kW

6 Which of the following statements is false?

    A   A d.c. motor converts electrical energy to mechanical energy.

    B   The efficiency of a d.c. motor is the ratio $\dfrac{\text{input power}}{\text{output power}} \times 100\%$.

    C   A d.c. generator converts mechanical energy to electrical energy.

    D   The efficiency of a d.c. generator is the ratio $\dfrac{\text{output power}}{\text{input power}} \times 100\%$.

7 The e.m.f. equation of a transformer of secondary turns N_2, magnetic flux density B_m, magnetic area of core a, and operating at frequency f is given by

    A $E_2 = 4.44 N_2 B_m a f$ volts

    B $E_2 = 4.44 \dfrac{N_2 B_m f}{a}$ volts

    C $E_2 = \dfrac{N_2 B_m f}{a}$ volts

    D $E_2 = 1.11 N_2 B_m a f$ volts

Questions 8 and 9 refer to the following information.

An 8-pole induction motor, when fed from a 60 Hz supply, experiences a 5% slip.

8 The synchronous speed is

    A   427.5 rev/min
    B   855 rev/min
    C   900 rev/min
    D   945 rev/min

9 The rotor speed is

    A   427.5 rev/min
    B   855 rev/min
    C   900 rev/min
    D   945 rev/min

10   In the auto-transformer shown, the current in section PQ is

A   1.6 A
B   1.7 A
C   5 A
D   3.3 A

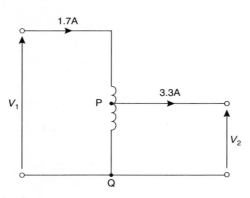

(for Test 12, Question 10)

11   Which of the following statements is false when referring to a three-phase induction motor?

A   The synchronous speed is half the supply frequency when it has four poles.
B   In a 2-pole machine, the synchronous speed is equal to the supply frequency.
C   If the number of poles is increased, the synchronous speed is reduced.
D   The synchronous speed is proportional to the number of poles.

12   A stator winding of an induction motor supplied from a three phase, 60 Hz system is required to produce a magnetic flux rotating at 900 rev/min. The number of poles is

A   2
B   8
C   6
D   4

13   The speed/torque characteristic shown is typical of

A   an induction motor
B   a series motor
C   a differential compound motor
D   a shunt motor

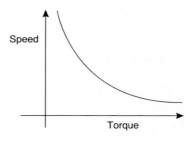

(for Test 12, Question 13)

14   The starting torque of a simple squirrel-cage motor

A   is high
B   increases as rotor current rises
C   decreases as rotor current rises
D   is low

15   Iron losses in a transformer are due to

A   eddy currents only
B   flux leakage
C   both eddy current and hysteresis losses
D   the resistance of the primary and secondary windings

16   The slip speed of an induction motor may be defined as

A   number of pairs of poles ÷ frequency
B   rotor speed − synchronous speed
C   rotor speed + synchronous speed
D   synchronous speed − rotor speed

17   With a d.c. motor, the starter resistor

A   limits the armature current to a safe starting value
B   controls the speed of the machine
C   prevents the field current flowing through and damaging the armature
D   limits the field current to a safe starting value

18 In d.c. generators iron losses are made up
    of

    A hysteresis and friction losses
    B hysteresis, eddy current and brush
       contact losses
    C hysteresis and eddy current losses
    D hysteresis, eddy current and copper
       losses

19 A step-up transformer has a turns ratio of
    10. If the output current is 5 A, the input
    current is

    A 50 A
    B 5 A
    C 2.5 A
    D 0.5 A

20 The speed of a d.c. motor may be
    increased by

    A increasing the armature current
    B decreasing the field current
    C decreasing the applied voltage
    D increasing the field current

Test paper 13 Electrical technology (2)

1 The phase voltage of a delta-connected three phase system with balanced loads is 240 V. The line voltage is

    A 720 V
    B 440 V
    C 340 V
    D 240 V

2 Which of the following statements is false?

    A A commutator is necessary as part of a d.c. motor to keep the armature rotating in the same direction.
    B A commutator is necessary as part of a d.c. generator to produce a unidirectional voltage at the terminals of the generator.
    C The field winding of a d.c. machine is housed in slots on the armature.
    D The brushes of a d.c. machine are usually made of carbon and do not rotate with the armature.

Questions 3 to 9 refer to the following information.

A 100 kVA, 250 V/10 kV, single phase transformer has a full-load copper loss of 800 W and an iron loss of 500 W. The primary winding contains 120 turns.

3 The total full-load losses are

    A 0.5 kW
    B 0.8 kW
    C 1.3 kW
    D 100 kW

4 The full-load output power at 0.8 power factor is

    A 0.4 kW
    B 0.64 kW
    C 20 kW
    D 80 kW

5 The full-load input power at 0.8 power factor is

    A 81.3 kW
    B 80 kW
    C 78.7 kW
    D 21.3 kW

6 The full-load efficiency at 0.8 power factor is

    A 80%
    B 81.30%
    C 98.38%
    D 98.40%

7 The half full-load copper loss is

    A 125 W
    B 200 W
    C 250 W
    D 400 W

8 The transformer efficiency at half full-load, 0.8 power factor is

    A 40%
    B 49.20%
    C 98.28%
    D 97.80%

9 The number of secondary winding turns is

    A 4800
    B 300
    C 48
    D 3

10 The input power to a three phase a.c. motor is 4 kW. If the voltage and current to the motor are 360 V and 7.2 A,

respectively, the power factor of the system is

A   0.891
B   0.648
C   0.617
D   0.356

11   A starter resistor is necessary with large d.c. motors to

A   increase the back e.m.f.
B   reduce the field current
C   increase the armature starting current
D   reduce the armature starting current

12   Which of the following statements is false?

A   For the same power, loads connected in delta have a higher line voltage and a smaller line current than loads connected in star.
B   When using a two-wattmeter method of power measurement the power factor is unity when the wattmeter readings are the same.
C   A.c. may be distributed using a single phase system with two wires, a three phase system with three wires or a three phase system with four wires.
D   The national standard phase sequence for a three phase supply is R, Y, B.

Questions 13 to 16 refer to the following information.
A three phase, 4-pole, 50 Hz induction motor runs at 1440 rev/min.

13   The synchronous speed is

A   1 rev/s
B   12.5 rev/s
C   24 rev/s
D   25 rev/s

14   The slip speed is

A   1 rev/s

B   12.5 rev/s
C   24 rev/s
D   25 rev/s

15   The percentage slip is

A   1%
B   4%
C   50%
D   96%

16   The frequency of induced e.m.f.s in the rotor is

A   0.05 Hz
B   2 Hz
C   25 Hz
D   50 Hz

17   Which of the following statements is false?

A   A series-wound motor has a large starting torque.
B   The speed of a series-wound motor drops considerably when load is applied.
C   A shunt-wound motor must be permanently connected to its load.
D   A shunt-wound motor is essentially a constant-speed machine.

18   The supply voltage to a d.c. motor is 240 V. If the back e.m.f. is 230 V and the armature resistance is 0.25 Ω, the armature current is

A   10 A
B   40 A
C   960 A
D   920 A

19   Which of the following statements about a three phase induction motor is false?

A   The speed of rotation of the magnetic field is called the synchronous speed.
B   The rotating magnetic field has a constant speed and constant magnitude.

C   A three phase supply connected to the rotor produces a rotating magnetic field.

D   It is essentially a constant speed machine.

20   The effect of inserting a resistance in series with the field winding of a shunt motor is to

A   increase the magnetic field

B   increase the speed of the motor

C   decrease the armature current

D   reduce the speed of the motor

Test paper 14 Electrical technology (3)

1 A 440 V/110 V transformer has 1000 turns on the primary winding. The number of turns on the secondary is

   A 550
   B 250
   C 4000
   D 25

Questions 2 and 3 refer to a three phase induction motor.
Which statements are false?

2 A As the rotor is loaded, the slip decreases.
   B The slip speed is the synchronous speed minus the rotor speed.
   C The frequency of induced rotor e.m.f.s increases with load on the rotor.
   D The torque on the rotor is due to the interaction of magnetic fields.

3 A If the rotor is running at synchronous speed, there is no torque on the rotor.
   B If the number of poles on the stator is doubled, the synchronous speed is halved.
   C At no-load, the rotor speed is very nearly equal to the synchronous speed.
   D The direction of rotation of the rotor is opposite to the direction of rotation of the magnetic field to give maximum current induced in the rotor bars.

4 A 4-wire three phase star-connected system has a line current of 10 A. The phase current is

   A 40 A

   B 30 A
   C 20 A
   D 10 A

5 A 4-pole, three phase induction motor has a synchronous speed of 25 rev/s. The frequency of the supply to the stator is

   A 25 Hz
   B 100 Hz
   C 50 Hz
   D 12.5 Hz

6 A transformer has 800 primary turns and 100 secondary turns. To obtain 40 V from the secondary winding the voltage applied to the primary winding must be

   A 5 V
   B 20 V
   C 2.5 V
   D 320 V

Questions 7 to 14 refer to the following information.
A shunt-wound d.c. machine is running at n rev/s and has a shaft torque of T Nm. The supply current is I A when connected to d.c. busbars of voltage V volts. The armature resistance of the machine is R_a ohms, the armature current is I_a A and the generated voltage is E volts.

7 The input power when running as a generator is given by

   A VI
   B $2\pi n T$
   C EI
   D $(E - I_a R_a)(I)$

8   The output power when running as a motor is given by

   A VI
   B $2\pi nT$
   C EI
   D $(E - I_a R_a)(I)$

9   The input power when running as a motor is given by

   A VI
   B $2\pi nT$
   C EI
   D $(E - I_a R_a)(I)$

10   The output power when running as a generator is given by

   A VI
   B $2\pi nT$
   C EI
   D $(E - I_a R_a)(I)$

11   The generated voltage when running as a motor is given by

   A $V - I_a R_a$
   B $E - I_a R_a$
   C $V + I_a R_a$
   D $E + I_a R_a$

12   The terminal voltage when running as a generator is given by

   A $V - I_a R_a$
   B $E - I_a R_a$
   C $V + I_a R_a$
   D $E + I_a R_a$

13   The generated voltage when running as a generator is given by

   A $V - I_a R_a$
   B $E - I_a R_a$
   C $V + I_a R_a$
   D $E + I_a R_a$

14   The terminal voltage when running as a motor is given by

   A $V - I_a R_a$
   B $E - I_a R_a$
   C $V + I_a R_a$
   D $E + I_a R_a$

15   The line voltage of a 4-wire three phase star-connected system is 11 kV. The phase voltage is

   A   19.05 kV
   B   11 kV
   C   6.35 kV
   D   7.78 kV

16   In the two-wattmeter method of measuring power in a balanced three phase system readings of P_1 and P_2 watts are obtained. The power factor may be determined from

   A $\sqrt{3}\left(\dfrac{P_1 + P_2}{P_1 - P_2}\right)$

   B $\dfrac{(P_1 - P_2)}{\sqrt{3}(P_1 + P_2)}$

   C $\sqrt{3}\left(\dfrac{P_1 - P_2}{P_1 + P_2}\right)$

   D $\dfrac{(P_1 + P_2)}{\sqrt{3}(P_1 - P_2)}$

17   The slip speed of an induction motor depends on

   A   armature current
   B   supply voltage
   C   eddy currents
   D   mechanical load

18   The expected characteristic for a shunt-wound d.c. generator is

   A   P
   B   Q
   C   R
   D   S

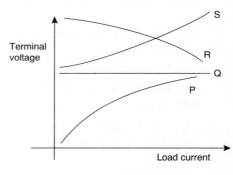

(for Test 14, Question 18)

19 A load is to be matched to an amplifier
    having an effective internal resistance of
    10 Ω via a coupling transformer having a
    turns ratio of 1 : 10. The value of the load
    resistance for maximum power transfer is

    A   1 kΩ
    B   100 Ω

    C   100 mΩ
    D   1 mΩ

20 The core of a transformer is laminated to
    A   limit hysteresis loss
    B   reduce the effects of eddy current loss
    C   reduce the inductance of the windings
    D   prevent eddy currents from occurring

Test paper 15 Electrical technology (4)

1 Which of the following statements is false for a series-wound d.c. motor?

  A The speed decreases with increase of resistance in the armature circuit.

  B The speed increases as the flux decreases.

  C The speed can be controlled by a diverter.

  D The speed can be controlled by a shunt field regulator.

2 The phase voltage of a 4-wire three phase star-connected system is 110 V. The line voltage is approximately

  A 440 V

  B 330 V

  C 191 V

  D 110 V

3 An advantage of an auto-transformer is that

  A copper loss is reduced

  B iron losses are reduced

  C it gives a high step-up ratio

  D it reduces capacitance between turns

4 The slip speed of an induction motor

  A is zero until the rotor moves and then rises slightly

  B is 100% until the rotor moves and then decreases slightly

  C is 100% until the rotor moves and then falls to a low value

  D is zero until the rotor moves and then rises to 100%

5 a 1 kV/250 V transformer has 500 turns on the secondary winding. The number of turns on the primary is

  A 2000

  B 125

  C 1000

  D 250

Questions 6 to 11 refer to the following information.

Three loads, each of resistance 16 Ω and inductive reactance 12 Ω, are connected in delta to a 400 V, three phase supply.

6 The phase impedance is

  A 4 Ω

  B 10.58 Ω

  C 20 Ω

  D 28 Ω

7 The line voltage is

  A 693 V

  B 400 V

  C 231 V

  D 11.55 V

8 The phase voltage is

  A 693 V

  B 400 V

  C 231 V

  D 11.55 V

9 The phase current is

  A 14.29 A

  B 20 A

  C 37.81 A

  D 100 A

10 The line current is

  A 11.55 A

  B 21.83 A

  C 24.75 A

  D 34.64 A

11 The total power dissipated in the load is

A   24 kW
B   19.2 kW
C   8 kW
D   6.4 kW

12 A commutator is a device fitted to a generator. Its function is

A   to prevent sparking when the load changes
B   to convert the a.c. generated into a d.c. output
C   to convey the current to and from the windings
D   to generate a direct current

13 The stator of a three phase, 2-pole induction motor is connected to a 50 Hz supply. The rotor runs at 2880 rev/min at full load. The slip is

A   4.17%
B   92%
C   4%
D   96%

14 Which of the following statements is false?

A   a transformer whose secondary current is greater than the primary current is a step-up transformer.
B   In a single phase transformer, the hysteresis loss is proportional to the area of the hysteresis loop.
C   In an ideal transformer, the volts per turn are constant for a given value of primary voltage.
D   In transformers, eddy-current loss is reduced by laminating the core.

15 The armature resistance of a d.c. motor is 0.5 Ω, the supply voltage is 200 V and the back e.m.f. is 196 V at full speed. The armature current is

A   4 A
B   392 A
C   400 A
D   8 A

16 Which of the following statements about a three phase squirrel-cage induction motor is false?

A   It has no external electrical connections to its rotor.
B   A three phase supply is connected to its stator.
C   A magnetic flux which alternates is produced.
D   It is cheap, robust and requires little or no skilled maintenance.

17 The power input to a mains transformer is 200 W. If the primary current is 2.5 A, the secondary voltage is 2 V and assuming no losses in the transformer, the turns ratio is

A   40:1 step down
B   40:1 step up
C   80:1 step down
D   80:1 step up

18 An ideal transformer has a turns ratio of 1:5 and is supplied at 200 V when the primary current is 3 A. Which of the following statements is false?

A   The turns ratio indicates a step-up transformer.
B   The secondary current is 15 A.
C   The transformer rating is 0.6 kVA.
D   The secondary voltage is 1 kV.

19 A 4-pole induction motor when supplied from a 50 Hz supply experiences a 5% slip. The rotor speed will be

A   25 rev/s
B   11.875 rev/s
C   26.25 rev/s
D   23.75 rev/s

20 The applied voltage to a d.c. motor is 120 V and the back e.m.f. is 116 V at full speed. If the armature resistance is 0.5 Ω, the armature current is

A   8 A
B   4 A
C   0.5 A
D   2 A

Answers

Test paper 1

1	C	2	B	3	C	4	D
5	B	6	B	7	D	8	B
9	D	10	A	11	A	12	C
13	D	14	A	15	C	16	A
17	A	18	D	19	B	20	C

Test paper 2

1	B	2	A	3	B	4	C
5	A	6	C	7	D	8	A
9	C	10	D	11	B	12	C
13	D	14	B	15	D	16	B
17	C	18	A	19	D	20	A

Test paper 3

1	C	2	D	3	C	4	B
5	A	6	D	7	A	8	B
9	C	10	B	11	A	12	D
13	B	14	A	15	C	16	D
17	B	18	D	19	C	20	A

Test paper 4

1	A	2	D	3	B	4	A
5	B	6	D	7	B	8	B
9	C	10	D	11	C	12	D
13	C	14	A	15	B	16	A
17	D	18	C	19	C	20	A

Test paper 5

1	C	2	B	3	A	4	C
5	D	6	A	7	C	8	D
9	B	10	D	11	C	12	B
13	D	14	C	15	A	16	B
17	A	18	D	19	B	20	A

Test paper 6

1	D	2	B	3	D	4	C
5	A	6	B	7	D	8	A
9	B	10	D	11	A	12	C
13	C	14	D	15	A	16	C
17	A	18	C	19	B	20	A

Test paper 7

1	C	2	D	3	C	4	A
5	D	6	B	7	C	8	C
9	D	10	A	11	A	12	B
13	C	14	A	15	B	16	D
17	B	18	A	19	D	20	B

Test paper 8

1	A	2	C	3	A	4	D
5	A	6	D	7	B	8	C
9	C	10	D	11	A	12	B
13	C	14	B	15	B	16	C
17	A	18	B	19	D	20	D

Test paper 9

1	D	2	B	3	A	4	D
5	D	6	C	7	A	8	C
9	A	10	B	11	A	12	C
13	D	14	B	15	D	16	B
17	C	18	A	19	B	20	C

Test paper 10

1	C	2	A	3	D	4	B
5	C	6	D	7	A	8	B
9	A	10	B	11	A	12	C
13	D	14	D	15	C	16	B
17	B	18	C	19	A	20	D

Test paper 11

1	C	2	D	3	B	4	B
5	D	6	B	7	B	8	A
9	C	10	A	11	C	12	A
13	B	14	C	15	D	16	D
17	A	18	A	19	C	20	D

Test paper 12

1	C	2	A	3	D	4	D
5	C	6	B	7	A	8	C
9	B	10	A	11	D	12	B
13	B	14	D	15	C	16	D
17	A	18	C	19	A	20	B

Test paper 13

1	D	2	C	3	C	4	D
5	A	6	D	7	B	8	C
9	A	10	A	11	D	12	A
13	D	14	A	15	B	16	B
17	C	18	B	19	C	20	B

Test paper 14

1	B	2	A	3	D	4	D
5	C	6	D	7	B	8	B
9	A	10	A	11	A	12	B
13	C	14	D	15	C	16	C
17	D	18	C	19	A	20	B

Test paper 15

1	D	2	C	3	A	4	C
5	A	6	C	7	B	8	B
9	B	10	D	11	B	12	D
13	C	14	A	15	D	16	C
17	A	18	B	19	D	20	A

ANSWER GRID

TEST PAPER NO. _____

	A	B	C	D
1	☐	☐	☐	☐
2	☐	☐	☐	☐
3	☐	☐	☐	☐
4	☐	☐	☐	☐
5	☐	☐	☐	☐
6	☐	☐	☐	☐
7	☐	☐	☐	☐
8	☐	☐	☐	☐
9	☐	☐	☐	☐
10	☐	☐	☐	☐
11	☐	☐	☐	☐
12	☐	☐	☐	☐
13	☐	☐	☐	☐
14	☐	☐	☐	☐
15	☐	☐	☐	☐
16	☐	☐	☐	☐
17	☐	☐	☐	☐
18	☐	☐	☐	☐
19	☐	☐	☐	☐
20	☐	☐	☐	☐

Name of student: _____

Class/group: _____ _____